Priscilla Hodder making nets at her cottage door at Eype, c. 1900. She is using a twine swifter, known also as a 'Bridport cross'.

ROPE,
NET ....ING

Anthony Sanctuary

Shire Publications Ltd

CONTENTS

Published by Shire Publications Ltd, Midland
House, West Way, Botley, Oxford OX2 0PH.
(Website: www.shirebooks.co.uk)
Copyright © 1980 and 1988 by the estate of
Anthony Sanctuary. First published 1980;
second edition 1988, reprinted 1996 and 2008.
Transferred to digital print on demand 2010.
Shire Library 51. ISBN-13: 978 0 85263 918 4.

Printed in Great Britain by PrintOnDemand-
Worldwide.com, Peterborough, UK.

ACKNOWLEDGEMENTS

Illustrations are acknowledged as follows: Bridport Museum, cover (photography by J. J. Pinn Ltd), pages 1, 6,
11, 12 (left); COI, page 3; from Pyne, *Microcosm*, page 12 (top); from Davis, *Fishing Gear of England and Wales*,
page 24 (top); Master Ropemakers Ltd, page 32. The publishers would like to thank Richard Sims for supplying
additional text for this reprint.

COVER: *Cottagers making nets (from the painting by Francis Newberry in Bridport Museum).*

BELOW: *A rope yarn spinner at work with hemp fibre round his waist. The big wheel is the
driving wheel for revolving hooks connected by pulley belt, rope and gearing. It revolves the
hooks at very high speed. The spinner walks backwards paying out fibre, which is caught on the
revolving hook and twisted into yarn.*

This retting tank, for soaking flax or hemp in order to rot away the pithy core, was photographed about 1946 at Slape Mills, Netherbury, near Bridport.

TRADITIONAL FIBRES AND THEIR MAN-MADE SUCCESSORS

Hemp (*Cannabis sativa*) has been the primary material for ropemaking throughout history. The hemp plant, a cultivated crop in most places, grows from 10 to 15 feet (3 to 4.5 m) in height and it is the supporting fibres under the skin of the stem which are used. The technical name is *bast* fibre. The crop is cut, soaked in water to rot away the pithy core and then beaten to separate the fibre from the stalk. The coarse texture of hemp fibre is ideal for ropemaking and allows for a good grip on the finished product.

Netmaking, by contrast, usually needs a softer and finer fibre and in past centuries flax (*Linum usitatissimum*) was preferred. Although the two plants are not botanically related, flax is rather like a miniature hemp and the crop is harvested in a similar manner, except that flax used to be pulled up by the roots rather than cut.

Since the late nineteenth century, however, some ropes have been made from Manila 'hemp' (*Musa textilis*) and sisal (*Agave sisalana*). Manila hemp comes from the leaves of a tree of the banana family which grows in the Far East. Sisal comes from the leaves of a spiky, fleshy plant grown in parts of central Africa. These are even rougher than hemp but are strong and suitable for heavy ropes and coarse twine. They came into fashion for the economic reason that they were cheaper to obtain.

Netmaking took advantage of the cotton industry and has used all types of cotton (*Gossypium*) from the 2 inch (50 mm) staple Sea Island to the $\frac{3}{4}$ inch (19 mm) staple Indian fibre. Cotton is soft, strong enough, reasonably water resistant and, by comparison, cheap.

The characteristics looked for in a

textile fibre, particularly for rope and net-making, are first tensile strength (wet and dry), then stretch and elasticity, abrasion resistance, rot and sunlight resistance and ability to withstand heat.

Strength is measured technically in grams per denier (a silk measure) but for simple comparison let us say that if a piece of best-quality hemp twine of a size suitable for tying a heavy parcel breaks at 100 pounds (45 kg), flax would be equally strong, Manila would hold 70 pounds (32 kg) and sisal and cotton about 50 pounds (23 kg).

Natural fibres are stronger when wet, have very little stretch, good abrasion resistance and can withstand sunlight and heat. They are all, however, subject to rot from bacterial action if allowed to lie wet for a prolonged period.

Man-made fibres were invented before the end of the nineteenth century, but nylon (1930s), polyethylene or polythene (1950s) and polypropylene (1960s) are the principal fibres used by rope and net makers today. They are all twice as strong as hemp and entirely resistant to rotting from bacteria. In contrast, they are subject to deterioration from sunlight and have comparatively low melting temperatures. Nylon loses a little strength when wet and all three have infinitely more stretch than the natural fibres. These man-made fibres are all derived from coal or oil and are, therefore, finite natural resources.

FIBRE	GROUP	APPEARANCE	STRENGTH Gr/denier	STRETCH %	SPECIFIC GRAVITY
FLAX	Bast	Hairlike	6.5	1.8	1.54
JUTE	Bast	Glossy	2.0	1.7	1.50
HEMP	Bast	Rough	6.5	1.8	1.48
SISAL	Leaf	Harsh	4.5	1.0	1.32
COIR	Seed	Springy	2.0	5.0	floats
COTTON	Seed	Fluffy	4.0	7.5	1.54
NYLON	Polyamide	Smooth	8.8	20.0	1.14
POLYTHENE	Polyolefin	Wiry	8.0	25.0	0.95
POLYPROPYLENE	Polyolefin	Glossy	8.0	20.0	0.91

The comparative properties of traditional and man-made fibres.

A lock of hemp represents a comfortable handful of fibres and is a convenient size for baling, transport and further processing.

A Dorset dozen of 36 pounds (16 kg) of yarn in its press in Bridport Museum. The twelve half-hanks total 21,600 yards (19,751 m) of yarn.

One of the less common uses of herring nets was to hold down the hay that covered the turf roofs of these fishermen's cottages at Cregneash in the Isle of Man.

THE HISTORY OF ROPE AND NET MAKING

The book of Genesis records that Noah built an ark. No kind of ship can be built and equipped without ropes. But ropes and nets began long before Noah. Monkeys twist vines and interlace them into primitive hammocks, so the craft is even older than man.

The earliest positive evidence the author has seen is a painting on the wall of an Egyptian tomb now in the Louvre Museum, Paris. This depicts fishermen hauling their net from the Nile in about 2600 BC. The men are shown, the ropes are shown and the fish are beautifully drawn, but the artist did not show the meshes of the net.

Trajan's Column in Rome is another source of information about ropes and nets. The detailed carvings of the legionaries show the reticulum or forage net carried on a forked stick. A charred fragment of one of these can be seen in the museum at Corbridge, Northumberland (Corstopitum). The specification of this net is identical with the horse's hay net of recent times.

The dark ages have left us with no authentic material, so we must move on to the record of William Mallet, Sheriff of Dorset and Somerset, whose pipe roll (account book) for 1211 reads: 'and for 1,000 ells of cloth for making ships' sails and for 3,000 weights of thread of hemp according to the measure of Bridport for making ships' cordage £48 9s 7d by the same writ. And for the expenses of Robert the fisherman who stayed at Bridport making his nets 39s by the same writ.'

The Bridport measure referred to is the Dorset *dozen* of yarn, which is still used. On a reel of 81 inches (2068 mm) circumference you wind eighty *bouts* (about the reel) and tie these with a knot. Ten knots make up a *half-hank* and twelve half-hanks make one dozen of yarn, a total of 21,600 yards (19,751 m). The weight of this bundle is the name of the yarn — 8 pounds, 24 pounds, 54 pounds and so on.

There is plenty of evidence of ropes for ships and agriculture and there are records of nets for coneys (rabbits), birds, thatched roofs and carts (to keep the pig secure). These reflect the principal occupations and means of livelihood of those days. Nets and ropes have always been part of daily life.

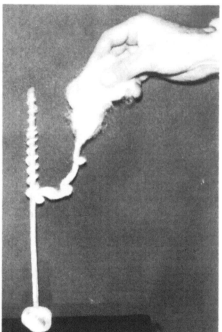

ABOVE: *A spindle whorl.* RIGHT: *A spinning walk illustrated in a mural in Bridport Town Hall.*

'Harry's broadsides' — King Henry VIII's big naval guns — had to be strongly roped in position, both to control them when fired and to stop them sliding at random in a storm and capsizing the ship.

About 1700 the naval dockyards at Chatham, Portsmouth and Devonport were coming into existence and so the making of the ropes for the Navy's ships was concentrated in local areas instead of in the traditional homes of ropemaking such as Bridport and Newcastle upon Tyne.

Nets were originally used for the essential elements of survival — hunting, roofing and bedding. As people found more time for leisure, nets were used more and more for sports and games. Royal (real) tennis, dating from Henry VIII's days and still played by many enthusiasts, needs a hazard net. Cricket (or *cricce* in medieval times) requires a practice net, as do discus and hammer throwing and golf.

The fishing fleets under sail used drift nets, spreading a wall of netting across 2 miles (3 km) of the North Sea, and when steam power gave the fishing vessels more strength they began to drag trawl nets along the sea bed.

Nowadays both fishing and sport have become specialised and scientific and so have the ropes and nets associated with them. The sports hall in a modern school or college has ropes and a climbing wall as well as nets which run out like curtains to provide for cricket, archery, netball, football, badminton and many other activities.

A twisting frame twists the yarns together before winding the twine on to a bobbin.

YARN AND TWINE

Yarn is a continuous series of overlapping fibres which have been twisted together so that the force of friction grips them and makes the yarn strong.

Originally yarns were spun on a *spindle whorl,* a weighted stick which was rotated, often by rolling along the thigh, in order to twist the fibres together. Then came the *spinning wheel* with its foot treadle, wheel, belt and flyer which enabled the spinner to keep the yarn twisting continuously. Yarns were also spun in *walks* from fibre wrapped around the waist. Modern machines still work on this principle of feeding fibres in at one end and twisting them together before winding the resulting yarn on to a bobbin.

Twine is two or more yarns twisted together. The method of manufacture is almost identical to that of spinning yarn except that you start with two or more twisted yarns instead of a series of loose fibres.

The theory of yarns, twines and ropes is that they are 'long things'. The traditional vegetable fibres used for yarns, twines and ropes are made up of cellulose cells, which are also 'long things', in turn made up of carbon 6 atoms, which in scientific diagrams are shown as strings of six beads. The ultimate fibre or cellulose cell of flax is 1⅛ inches (29 mm) long and 1/1000 of an inch (0.025 mm) in diameter.

Yarns, twines and ropes can be made by machine nowadays, but the ropemakers of older days were accustomed to making all these in a walk. The principle of the walk is that yarns are stretched out between revolving hooks, often 300 yards (275 m) apart, and these hooks twist the yarns together. So that the yarns do not sag on to the ground, they are supported by horizontal bars called *skirders* every few yards and these skirders have vertical pegs on them to separate the yarns.

The layout of a medieval town like Bridport lends itself well to family rope or twine walks because of the long narrow alleys which stretch back from the main streets. While his wife and daughter were making nets in the kitchen, a man would make twine and small ropes in the alleys off the main street with the help of his son, who turned a wheel to revolve the hooks. For many years after the introduction of local factories the walk method continued: indeed the last Bridport walk closed down only in 1970.

9

These ropemaker's tops are used to keep separate the groups of yarns that make up a rope while they are being twisted into strands. The top with four grooves is for making four-strand rope, used for shrouds on sailing ships and for rope ladders, where the rope has to be equally divided. The double top is for making two fishing lines at a time.

MAKING A ROPE

At one end of the rope walk is the *jack*. This is a frame at waist height fitted with three hooks pointing along the walk. The hooks are rotated by a crank handle and either gear wheels or a driving wheel and pulley belt. At the *lower end* of the walk is a freely swivelling hook which can be drawn part way along the walk as the rope contracts during the making.

The amount of yarn necessary to make the rope is now paid out. The more yarns, the bulkier the rope will be. Let us suppose that six yarns are tied on to each of the three hooks on the jack and their other ends are hitched to the swivelling hook at the lower end. Skirders must be placed all the way along the walk, otherwise the yarns will drag on the floor and each group will get tangled with the next.

Now the work begins. The ropemaker takes his *top*, a conical wooden block with three grooves to take the three groups of yarn, and places it close to the swivelling hook at the lower end with a group of yarns in each groove. This stops the yarns getting tangled as the twisting begins. The ropemaker's assistant starts to turn the crank handle of the jack and the hooks revolve at high speed. Each group of yarns starts to twist into a strand. The assistant keeps turning the handle. Soon the strands begin to shrink because of the twist. The ropemaker has to move slowly along the walk. When the strands have contracted by about a quarter of the length of the walk they are as highly twisted as is possible without kinking.

Now the second stage begins. The assistant keeps on turning his crank handle but the ropemaker walks steadily along the walk and then the magic happens! Because of all the twist in the strands, the swivelling hook at the lower end is forced to rotate and the rope makes itself behind the ropemaker as he walks forward. This makes a medium or soft laid rope. In order to make a hard laid rope, the swivelling hook must be positively driven and the ropemaker must walk more slowly.

Some ropewalks changed to steam and later to electrical power to improve production. The ropewalk at Chatham Dockyard was operated by HM Navy until 1983. It still produces rope using traditional methods. Modern machines, however, turn out rope by means of complex gearing, hoops and spindles.

Here in Bridport Museum a rope is being made. The ropemaker holds the top and as he moves forward the rope is formed behind him. His assistant keeps turning the crank handle of the jack, revolving the three hooks to which the three groups of yarn are attached.

Ropemaking at the beginning of the nineteenth century.

LEFT: *The lower end of a twine walk, with weight and pulley to allow for the contraction of the twine. As the jack at the other end of the walk turns and twists the twine, the weight rises and allows the hook holding the twine to move towards the jack.*

BELOW: *This rope jack has five hooks to twist yarn for five-strand rope, although normally only three are used.*

Ropemaking by machine at the walks of Rendall and Coombs in Bridport in the 1950s. Below, a laider ties yarns on to a mechanically driven jack.

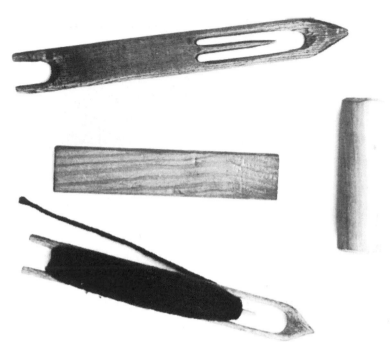

Net braiding needles (top and bottom), and measuring sticks or laces (centre).

NETMAKING BY HAND

Netmaking has always been a hand skill and remains so today.

The main kinds of knot used are the sheet bend (fisherman's method), the sheet bend (Bridport or small-mesh method), the double sheet bend and the reef knot. Other patterns tend to create a design rather than a plain row of meshes and are more allied to macrame work.

The *sheet bend,* originally used to attach a rope to the loop at the corner of a sail, consists of passing the rope through the loop and throwing a hitch round the lot. This is the fisherman's method and the tools required to speed the work are the *braiding needle,* so designed that it carries the maximum length of twine and is still narrow enough to go through the mesh, and the *lace* or measuring stick. The term 'lace' is a memory of the days when lace net was knotted rather than woven.

The needles may be of wood, metal or plastic. Modern ones are usually made of nylon. They may be of the pronged type or they may be double-ended.

The lace may be flat, round or of elliptical section, depending on the method of knotting being used. While the fisherman uses a flat measuring stick for his method, the Bridport knot, especially suited to small-mesh work because the needle passes through a whole mesh instead of only half a mesh, needs a round-section lace.

The *Bridport knot* is still a sheet bend, but it is made round the fingers of the left hand and the needle is inserted as a single movement, in contrast with the two stages of forming the fisherman's knot. The effect is to make the knot lie on its side and the resulting twist helps to eliminate knot slippage. Claims as to which method

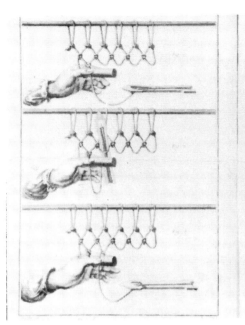
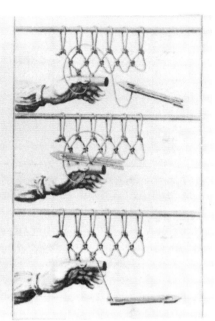

ABOVE: *Hand braiding by the fisherman's method, in which a needle is passed up through the previous mesh and a hitch is thrown around that mesh.*
BELOW: *The Bridport method of hand braiding, in which a loose slip knot is formed round the fingers of the left hand and the needle is passed through the sliding knot. (This is the method used by net machines.)*

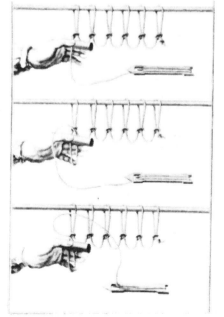
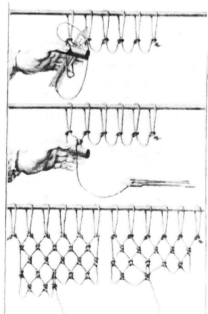

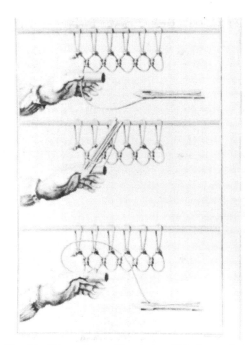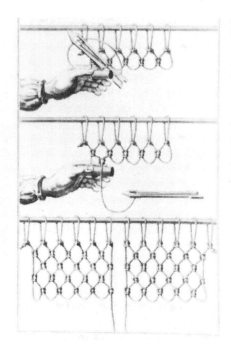

Hand braiding using reef knots, in which the needle is passed down through the previous mesh and a hitch is thrown round that mesh, finishing down again through the mesh.

is the quicker have yet to be settled by expert ladies in the presence of a timekeeper.

The *double knot* is used to reduce knot slippage when working with slippery man-made fibres such as nylon and it is merely the fisherman's method with an extra turn in the hitch.

Reef knots are used in the Far East and were used in Roman times. In simplest terms, the braider passes the needle down through the mesh instead of upwards before throwing a hitch.

Netmaking by hand is very similar to knitting. You can 'cast on' the chosen number of meshes, and you can increase or decrease by braiding twice into the edge mesh or by braiding two meshes together.

To make square-mesh netting, such as in a tennis net, the maker starts at one corner with two meshes and increases at the end of each row until the maximum width has been achieved. To produce a rectangle, she increases on one side and decreases on the other, and to taper away to nothing she decreases at both ends until there are only two meshes left to be tied off.

Hand making of nets survives as the best way of manufacturing small and awkward-shaped nets such as billiard-table pockets and magazine racks for aircraft. The great sheets of netting required for cutting into fishing nets, cricket nets and camouflage nets for the armed forces must be woven on looms and that is the subject of the next chapter.

Operating the treadle on a Jumper machine that dates from 1860. The photograph was taken about 1966 in the Joseph Gundry factory. The machine is now in Bridport Museum.

NETMAKING BY MACHINE

The first netmaking machine, invented in 1812 and put into practical use in the 1820s, was the *Jumper*, so called because it required two hundred or more wire springs to be compressed by a foot-operated treadle on which the operator jumped.

The principle of machine netmaking is to form a slip knot, pass a separate end of twine through the sliding loop and draw tight, thus capsizing the knot and forming a sheet bend. All netmaking machines work on this principle even today — some doing it horizontally, some vertically, but always by capsizing a slip knot.

In the 1890s M. Zang of Paris invented the multi-shuttle machine, on which slip knots were passed over many weft ends at the front of the machine rather than running a new thread through the slip knots wrapped around hooks and needles as on the Jumper.

There were power-driven forms of the Jumper designed to make sheets of herring net for the North Sea. They were made by de Sereville in France and commonly known as *Mons* machines. These tied up to one thousand knots at each cycle of the machine and there was a double knotting version which had no less than three camshafts behind it, an engineering nightmare. When first introduced to nylon, the double-knot Mons threw off sparks of static electricity until fitted with 'lightning conductors'.

The Zang machine served netmakers for nearly fifty years until the Japanese devised a new form of machine during the Second World War. A high-quality version

Netmaking machines work on the principle of capsizing a slip knot.

of this is among the most successful net machines of today. In the interval there have been improved versions of Zang's invention, and his grandson worked with a Finnish engineer named Ohls to produce another successful loom in the late 1960s. British technicians developed the ideas of Zang, but only to the extent of the Porlester machine, which had limited success for making heavy nets for the trawler fishermen working out of Hull, Grimsby and Fleetwood.

The development of netting looms has depended on their speed of operation – first five hundred knots per minute with one operator, then, because automatic stop motions enabled one operator to supervise two or more looms, two thousand knots per minute per person. Economy of personnel was vital because netmaking takes place in small communities, not in the big industrial cities.

Netmaking looms are not designed to make fancy shapes. They turn out sheets of 'cloth' which then have to be tailored into nets for industry, sport, fishing and other uses and these nets may be of any texture, weight or shape. This is the reason for the continuing importance of handwork in netmaking. The loom may be able to produce a sheet of cloth but only human fingers can gather the loops into position for selvedging edges and fitting ropes, floats, sinkers, poles and other gadgetry.

A Mons-type netting machine from the 1870s, still in use in the 1950s.

An early Zang-type netting machine from 1920s (above) and (below) the new Zang Lindemann netting loom developed in the 1960s.

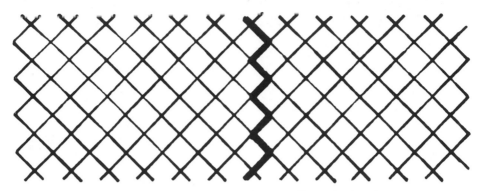

ABOVE: *Diamond mesh joined by cross join to produce an endless strip of netting.*

BELOW: *The three dimensions of netting.*

square 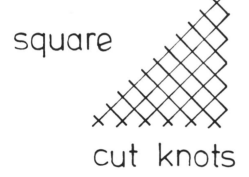 clean

loops

cut knots

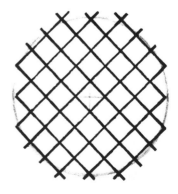

RIGHT: *Plan for cutting a circular piece.*

Selvedging a square edge.

NET FITTING

Our 'cloth', the sheet of netting, is made. Next it must be tailored. This process is referred to as net fitting.

The stages are usually: (1) cut the sheet; (2) join the major pieces; (3) cut the individual article; (4) selvedge the final net.

Netting comes from the loom as a diamond pattern and may be cut horizontally across the sheet, thus retaining the diamond mesh, or it may be cut diagonally in order to produce a square-mesh selvedge as in a tennis net. The side edges of the sheet are *clean loops* while the leading edge has *cut knots*. Thus we have the 'three dimensions of netting'.

The clean loops can be joined invisibly by making half a mesh between them so that it is possible to have an endless strip of netting of any depth, whether diamond or square mesh at the edge. Joining the clean loops is known as a *shut* or *cross join;*

joining the cut knots is known as a *scun* or *long join*. It is also possible to join square mesh edges by a zigzag process.

Next comes the cutting of the piece. This may be straight-edged or tapered or a combination of the two. As with the decrease (shrink) and increase (gain) in hand braiding, so the edge of a piece of machine-made netting can be shaped at will. Given a reasonable number of meshes, it is simple to cut out a circular piece of netting.

Having described the cutting and joining, mending must be referred to. Whether because of a fault on the loom or because of damage in use, nets often have to be mended. This implies cutting out a suitable hole so that the repair can follow the dimensions of netting and be almost invisible.

The final process is selvedging. Just as a

piece of cloth must be hemmed to stop the weave from fraying or to strengthen the edge, so must a net be selvedged to prevent the cut knots from pulling apart or to give added strength. Selvedging the clean loops is only necessary for strengthening, but this is frequently the part of a fishing net that gets the roughest handling. Selvedging the cut knots is a wise precaution in case some of them pull apart, and selvedging a square edge is vital because the edge is usually under strain when in use and liable to come apart.

Depending on the type of net and the intended use, it is often cheaper and more convenient to use a sewing machine or even a web binding along a square edge. This cannot be done with diamond mesh since netting is an infinitely flexible fabric — you can wrap a square of netting round a football and it will fit — and only the fingers can pick up and control the loops of the mesh.

Rigging a football net to fit the shape of the goalposts and back supporting poles.

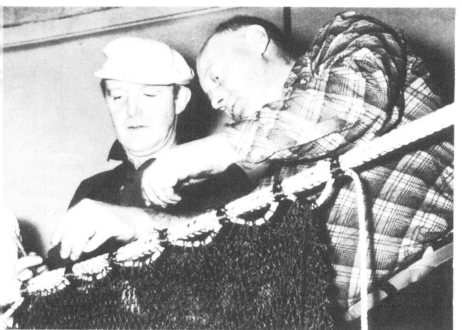

Roping a purse seine, which is a fishing net for catching herring, pilchard and sometimes even cod and tuna. The net, often a quarter of a mile (400 m) long and 100 yards ((90 m) deep, is used to encircle a shoal. The bottom is closed with a draw rope and the fish are scooped into the vessel.

NET RIGGING

The very flexibility of netting makes it necessary for most finished nets to be attached to a frame. This may be a solid structure like a goalpost or a fruit cage, but frequently the frame is of rope.

The classic example is the Scotch herring net. As an understanding of Latin gives you the roots of many languages, so an understanding of the Scotch net enables you to rig most other kinds. The Scotch net is made from fine cotton. It has a few meshes of heavier material as a border; it has single cords at each side; it has double cords at each end; it has tiers to attach it to its ropes, which may be single, double or quadruple and fitted with a variety of corks, leads, grommets and identification marks.

Rigging, therefore, is the attachment of a net to its frame. The simplest method is to run a rope through the edge meshes, but this may not be rigid enough. The rope can be fixed to the net every few meshes or even at every mesh. Attachment may be by tying alternately a knot into the mesh and a hitch round the rope; it can be by threading a length of twine through one or more meshes and then making a clove hitch, a marling hitch or a rolling hitch on the rope. Square-mesh netting can be lashed to the rope at every knot or it can be hung by alternate loops known as flyers or flying meshes.

The secret of rigging lies in knowing when a piece of netting should be stretched taut and when it should be free to 'flow' in the water or at the shock of a ball striking it. The slope at the back of a football goalnet is a precaution against the ball bouncing out of the goal. The slackness in a golf practice net is to take the shock of the ball striking it. The rigidity of a tennis net is a deliberate and identifiable hazard. The slackness in the corners of a herring net is to ease the strain of the warp in rough weather. The various adjustments along the headrope of a trawl net are to spread the strain of

23

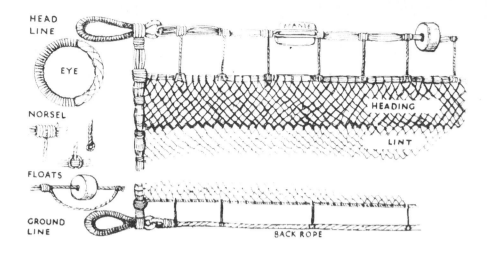

HEAD LINE

EYE

NORSEL

FLOATS

GROUND LINE

HEADING

LINT

BACK ROPE

ABOVE: *The construction of a Scotch herring drift net.*
BELOW: *Fitting floats on a fishing net to ensure buoyancy, either to make the net float on the surface or to keep it upright in deep water.*

towing evenly across all the meshes.

Fishing nets are equipped with floats, normally plastic nowadays rather than cork, and sinkers, which may be pieces of lead or chain or even lead pellets plaited into a cord. The number and spacing depends on whether the net is for towing or stationary and whether it is to float near the surface or lie at the sea bed.

Sports nets may be divided into hazard nets, normally stretched between posts, goalnets, normally on a frame, whether it be for football or netball, and enclosures, as for cricket practice or a tennis court surround, when the support may be poles with guyropes or a more sophisticated frame. Sometimes the net is lashed to its frame, sometimes attached by hooks. There may be supporting wires, canvas binding, tie cords and all manner of other gadgetry, but until the net is rigged to its supporting frame it cannot be said to be complete.

Measuring a fishing net. The rigger on the left carries a yard stick. Smaller distances are judged by eye or by hand span.

ABOVE: *The old line walks at North Mills in Bridport are now demolished. The walks were inside and ran the length of the building, with a net fitting loft above them.*
BELOW: *A row of typical netmakers' cottages in South Street, Bridport.*

The cart from Pymore Mill delivered twine and collected netting from the outworkers. It survived until 1950, the date of this photograph.

LIFE IN
NINETEENTH-CENTURY
BRIDPORT

The day started early, as in other parts of industrial Britain. The man of the house had to be in the mill or ropewalk at 6 a.m. and by the flourishing period of the 1870s the children were attending half day school before going to work.

A team of ropers (ropes) or laiders (fishing lines) consisted of the laider himself, who made the lines, together with the top-end man and the lower-end man, who turned the hooks.

Hemp was the material for bigger lines and ropes, while flax was used for twine and canvas. The best-quality hemp came from Italy, white in colour, long in the fibre and silky to the touch. Russian hemp from the Baltic ports was dark, short and rough, but equally strong.

Bridport was then prosperous from the cod-fishing lines for the Grand Banks of Newfoundland. People still living have heard their grandfathers reminiscing about that period and those grandfathers carried their own grandfathers' memories dating back to 1800.

There is a record of a ship of 132 tons (134 t) burthen sailing from Bridport Harbour to Newfoundland in 1887, doubtless full of fishing lines, and the mill owners, the Gundrys, the Hounsells, the Rendalls, the Stephenses and the Whethams, were employing the working members of Bridport's 6,700 population to the full.

The great ropes for ships' hawsers were made by heavy machinery in Newcastle upon Tyne, Gourock on the Clyde and Belfast, but the great variety of medium-sized cordage was the speciality of Bridport.

Wages have always been a touchy subject and memories of being paid in pieces of salt cod would have survived into the 1870s, although by then a team of laiders could earn 50 shillings a week between them.

Conditions in a ropewalk were always rough. It was a long shed, frequently with no glazed windows, freezing cold in winter and swelteringly hot in the summer. A brazier at each end for the men to warm their hands was the only practical measure possible in winter.

Those living on the fringe of the town had to walk 2 miles in order to purchase the household necessaries, although each cottage had its vegetable patch and possibly a goat and a few chickens.

Even in more recent times the braiders working the Jumper looms would start at six o'clock, taking a pint of ale on their way to work. They would stop for breakfast and another pint at about eight, again for lunch at ten and, if they lived sufficiently nearby, they would go home for dinner at noon. They were paid by the *piece* — 50 yards (45.7 m) of net — and to that extent were their own masters about timekeeping. The pattern for the afternoon was similar.

Once the housewife had sent the children off to morning school, she could concentrate on the household chores and the preparation of the midday dinner. In the afternoon she and her mother would settle down to net braiding over the kitchen table and this was often a social occasion when neighbours came to join in.

The youngest children filled the braiding needles from the *ran* or hank of twine stretched on the swifter known as the *Bridport cross*. Each of the villages on the outskirts of Bridport specialised in coarse or fine work, large mesh or small mesh, according to tradition.

Those living close to the centre of the town would collect their twine personally, receiving payment as they delivered their finished work. In the villages the braiders relied on the carrier's cart to bring twine to them, but some who were eager to earn more money would walk 3 or 4 miles (5-6 km) into town to fetch more twine.

Payment was not for the individual article, but for the length of twine converted into netting. A tennis net, 14 yards (13 m) long and twenty-eight meshes deep, requires approximately a ran of twine — 766 yards (700 m) and almost eight thousand knots. At this time the payment would have been 15 old pence for about eight hours work.

A century later, after two world wars and the introduction of the internal combustion engine, the people of West Dorset still work at the braiding of nets over the kitchen table and the money they receive, although infinitely more in cash, is still pin money rather than a means of livelihood.

Delivering by van to outworkers living at Loders, 2 miles (3 km) east of Bridport.

THE NETMAKING
INDUSTRY IN BRIDPORT

The netmaking trade of modern Bridport owes its origin to Samuel Gundry in 1665. He was not a factory owner in the modern sense, but a man of substance, a banker, a merchant and owner of the building known today as The Court, which is the present office and headquarters of the internationally famous company which carries on the Bridport trade. Samuel Gundry was able to purchase a large part of the local hemp crop, arrange for its conversion into yarns, twines and nets and then market the finished product.

By 1670 William Hounsell had set up an organisation for spinning yarns, twisting twines and making ropes, and in the eighteenth century the names of Rendall, Whetham and the Pymore Mill were well known.

Netmaking was, until comparatively recently, a hand craft. There was no need for the resources of water power and coal necessary for the cotton and wool spinning of the industrial revolution. This is why many of the early factories at Bridport were scattered around the town.

Ships from Bridport Harbour, now known as West Bay, provided sea transport to markets in Europe, Africa and the Americas for fishing nets, although the nets for agriculture and catching birds and game were marketed more locally and more easily because of their smaller bulk.

The output of Pymore Mill in 1829 was 7648 hundredweights (389 t), valued at £5250, and this was probably matched by each of the other mills, Hounsells in Bridport and Rendalls at Burton Bradstock. It could be assumed that one third of this was destined for netting, one third

An outworker hand-braiding heavy plaited nylon cord using a tubular lace. These nets are for containing air freight or for use by helicopters.

for canvas and one third for cordage. It was at this period that the first netmaking loom, the Jumper, was invented. Originating in Scotland, it was soon copied by the Bridport engineers.

A hard working family could save up its money and purchase a Jumper loom, which was usually housed in a lean-to-shed behind the cottage. Before long the investment paid off and a second loom was acquired and a hired man taken on to operate it. This was how most of the smaller netmaking firms began.

At the beginning of the twentieth century there were fifteen family firms manufacturing yarn, twine, rope and nets. As the years went by, some firms faded out, some were absorbed into stronger ones and some amalgamated until, at the end of the Second World War, there were two groups, the Bridport Industries, comprising Edwards, Gales, Hounsells, James and Rendall & Coombs, and the Joseph Gundry Group including Pymore Mill, Tucker and Whetham.

War has always brought heavy demand for Bridport products, including cordage for the armed services, pull-throughs for cleaning rifles, nets for camouflage, anti-submarine booms, helmet covers and all kinds of container from those carrying victuals to those for hoisting lorries on board ship. These were the principal products in 1914-18 as well as in 1939-45 and many of the modern factory buildings grew out of the proceeds. Between the world wars, Bridport struggled through slump conditions, as did most of the textile trade, but the increasing popularity of sports such as tennis, cricket and football resulted in a great call for nets at a crucial time.

In the early 1950s surplus stores from the Army were converted for civilian use. Camouflage nets bought at auction for a song were stripped and resold for the growing of peas and beans. Tons of smaller cordage, no longer needed by the Navy and the Air Force, were to be found in ironmongers' shops, and so the demand for new products was much reduced. Even when these bargains had been exhausted, the makers had the problem of replacing old machinery. Manufacture of modern netmaking looms of the Zang type had not yet been resumed in France and an attempt was made to motorise the old Jumper looms. The Ackerman family worked wonders at this, and the Brussels sprout growers of Evesham and thousands

ABOVE: *Fitting a football goalnet.*
BELOW: *Roping a climbing net in heavy sisal. The rope is too thick to wind on a needle and often the men would wrap their hands in strips of motor-tyre inner tubing for protection against the rough rope.*

of badminton players throughout the world would have been without their nets had it not been so.

The Bridport netmaking groups spent ten years in cut-throat competition with each other until, in 1963, through the efforts of a 'foreign' managing director from Somerset, compromise was achieved and the Bridport-Gundry Group was set up.

Rationalisation followed, and manufacture concentrated on just three sites in Bridport. Still bound to its fishing base, Bridport Gundry set up depots in the UK and Ireland, but the traditional market was in decline and alternatives were needed.

Exploiting their links in the aviation industry, Gundry manufactured cargo restraint nets under licence from Hawker Siddeley from 1971. In 1974 Geoff Dilbey, B G's chief design engineer, invented the breakthrough 'knotless net', which made nets lighter and cheaper to make. Bridport Aviation Products was launched in 1979 to develop products for the aviation and defence industries, and became the dominant force within the company.

Life grew harder for traditional netmakers, so Bridport Gundry Marine moved into trawlmaking, selling completed products rather than nets alone. The continued decline in the fishing market led to the company pulling out of North America and selling its Scottish and Irish operations in the early 1990s.

In 1993 tough new targets were set for the trading divisions, which only the aviation/defence and medical divisions could meet. As a result the other, traditional companies were sold in 1997.

In 1999 Bridport plc was sold to the Marmon Group, remaining part of that concern until 2004 when it became part of AmSafe Partners Inc. Now known as AmSafe Bridport, the company remains a world leader in aviation restraint systems, with bases around the world.

Edwards Sports Products, one of the companies sold in 1997, is run as a sports equipment company and provides complete sports systems. Edwards continues to produce tennis equipment and football nets and will be supplying the Beijing Olympics.

Nine other netmaking companies remain in Bridport, ranging from internationals such as Huck UK and Sicor International UK, to smaller family concerns, such as Knowle Nets and Coastal Nets. They produce nets for sports and play areas, safety netting, garden nets and game nets – fishing nets are now a small part of the output. In all cases the sheet netting is imported and assembled into the finished product in factories and workshops in Bridport. Hand braiding continues to the present day, supplying P J Winchester of Bishops Lydeard, who recently bought Redport Nets.

PLACES TO VISIT

Bridport Museum, The Coach House, Gundry Lane, Bridport, Dorset DT6 3RJ. Telephone: 01308 422116. Website: www.bridportmuseum.co.uk Archive material relating to the trades of rope and net making.

Chatham Historic Dockyard, Chatham, Kent ME4 4TZ. Telephone: 01634 823807. Website: www.chdt.org.uk Demonstrations of rope making at the ropery.

W. R. Outhwaite & Son, Town Foot, Hawes, North Yorkshire DL8 3NT. Telephone: 01969 667487. Website: www.ropemakers.com Traditional rope making can be watched free of charge on Monday to Friday all year, except Christmas, and Saturdays, July to October. Educational visits by appointment.

OTHER MUSEUMS WITH ROPE OR NET MAKING EXHIBITS.

Bewdley Museum, The Shambles, Load Street, Bewdley, Worcestershire DY12 2AE. Telephone: 01299 403573. Website: www.wyreforestdc.gov.uk

Fleetwood Museum, Queen's Terrace, Fleetwood, Lancashire FY7 6BT. Telephone: 01253 876621. Website: www.fleetwoodmuseum.co.uk

Gairloch Heritage Museum, Achtercairn, Gairloch, Ross-shire IV21 2BP. Telephone: 01445 712287. Website: www.gairlochheritagemuseum.org.uk

Michelham Priory, Upper Dicker, Hailsham, East Sussex BN27 3QS. Telephone: 01323 844224. Website: www.sussexpast.co.uk

Milford Haven Heritage and Maritime Museum, The Old Custom House, The Docks, Milford Haven, Pembrokeshire SA73 3AF. Telephone: 01646 694496. Website: www.visitpembrokeshire.com

National Fishing Heritage Centre, Alexandra Dock, Great Grimsby, Lincolnshire DN31 1UZ. Telephone: 01472 323345. Website: www.nelincs.gov.uk/leisure/museums/FHC.htm

Ropewalk Contemporary Arts and Crafts, The Ropewalk, Maltkiln Lane, Barton upon Humber, North Lincolnshire DN18 5JT. Telephone: 01652 660380. Website: www.the-ropewalk.co.uk